# Blue Whales
## and Buttercups

by Megan Goss, Jonathan Curley,
and Ashley Chase

Published and Distributed by

**Published and Distributed by**

 These materials are based upon work partially supported by the National Science Foundation under grant number ESI-0242733. The Federal Government has certain rights in this material. Any opinions, findings, and conclusions or recommendations expressed in this material are those of the author(s) and do not necessarily reflect the views of the National Science Foundation.

 © 2009 by The Regents of the University of California. All rights reserved. No part of this publication may be reproduced or transmitted in any form or by any means, electronic or mechanical, including photocopy, recording, or any information storage or retrieval system, without permission in writing from the publisher.

Developed at Lawrence Hall of Science and the Graduate School of Education at the University of California at Berkeley

Seeds of Science/Roots of Reading® is a collaboration of a science team led by Jacqueline Barber and a literacy team led by P. David Pearson and Gina Cervetti.

Delta Education LLC
PO Box 3000
Nashua, NH 03061
1-800-258-1302
www.deltaeducation.com

Blue Whales and Buttercups
594-0023
ISBN: 978-1-59821-497-0
1 2 3 4 5 6 7 8 9 10  14 13 12 11 10 09

# Contents

**How Living Things Are Different** . . . . . . . . . . . . . . . .4
   Living Things Grow to Different Sizes . . . . . . . . . .6
   Living Things Get Around in Different Ways . . . . . .8
   Living Things Protect Themselves
      in Different Ways. . . . . . . . . . . . . . . . . . . . . .10

**How Living Things Are Similar** . . . . . . . . . . . . . . . . .12
   Some Living Things Have Four Limbs. . . . . . . . . . .14
   Some Living Things Have Feathers . . . . . . . . . . .16
   Some Living Things Have Flowers . . . . . . . . . . . .18

**All Living Things Are Related** . . . . . . . . . . . . . . . . . .20

**Glossary** . . . . . . . . . . . . . . . . . . . . . . . . . . . . . . . .24

# How Living Things Are Different

There are so many different kinds of animals, plants, and other living things on Earth. There are frogs and ants and roses and trees and sea stars and snakes.

Living things can have very different **characteristics**. A characteristic is anything you can notice about the way a living thing looks or acts. Some animals have fur, and others have feathers. Some plants have flowers, and others do not. Some animals protect themselves by running fast, and others protect themselves by biting. We call these differences in living things **variation**.

In order to study variation in living things, scientists collect information about plant and animal **species**. Scientists **observe** how animals or plants look on the outside and on the inside, too. They also observe what living things do. Each species has its own characteristics that make it different from other species.

# Living Things Grow to Different Sizes

You can find lots of variation in the sizes of living things. **Compared** to people, elephants are huge. But it would take about 20 big elephants to match the weight of just one blue whale. And the biggest trees weigh more than 40 blue whales!

Even living things that share many characteristics can be very different in size. When the smallest bats spread their wings, they only stretch about six inches wide. The biggest bats can spread their wings about 60 inches (five feet) wide. That's ten times wider!

17 feet

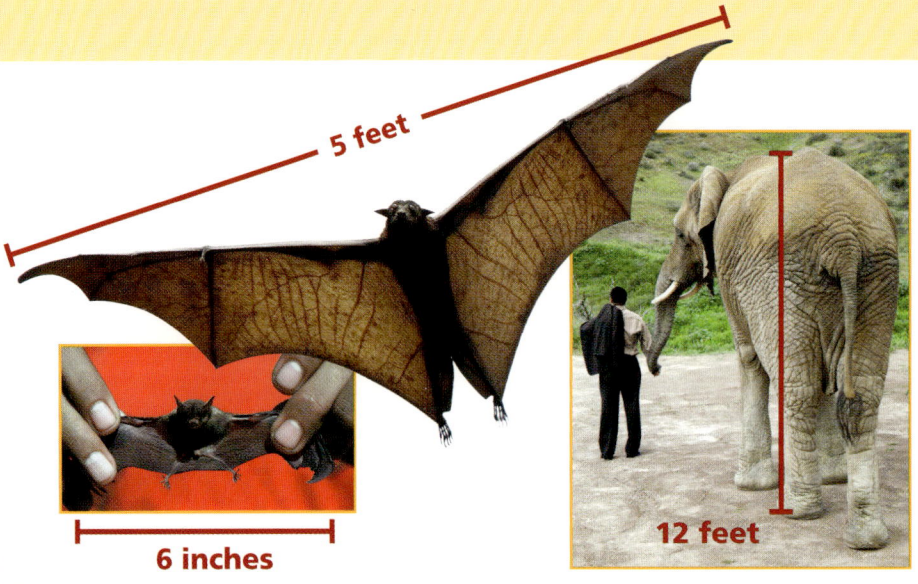

5 feet

12 feet

6 inches

9 inches

100 feet

Blue whales are the largest animals in the world.

6

Below are tiny plant-like living things that float in the ocean. They are too small to see without a **microscope**.

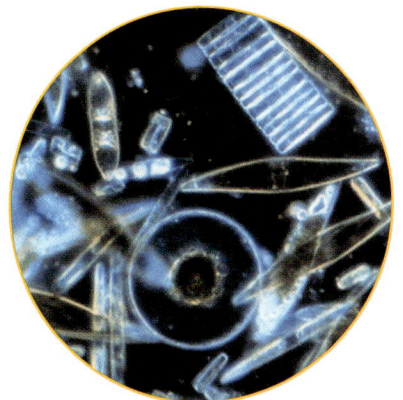

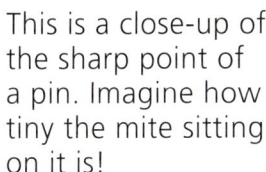

This is a close-up of the sharp point of a pin. Imagine how tiny the mite sitting on it is!

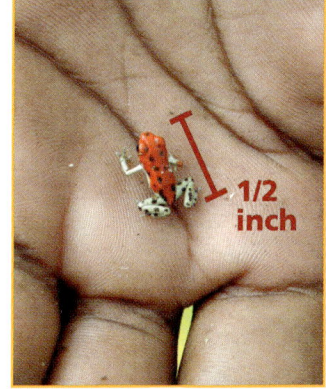

1/2 inch

3 feet

Giant sequoia trees are the world's largest living things.

200 feet

But difference in size is not the only kind of variation among living things on Earth . . .

**7**

# Living Things Get Around in Different Ways

All animals move, but different animals often move very differently. Animals may walk, run, climb, fly, swim, or slide. Plants don't move around the way animals do, but their seeds get from place to place in different ways. We can see lots of variation in the many ways that living things get around.

This snake slides across the sand by moving its body from side to side.

This sea anemone stays stuck in one place for most of its life. Still, it moves its tentacles around to catch food.

An octopus moves by squirting water out of its body. The squirt sends the octopus shooting through the ocean.

Burrs are seeds covered with tiny hooks. They get stuck to animals and carried to new places.

Coconuts are a kind of seed. They float from island to island.

Living things are different in other ways, too . . .

**9**

# Living Things Protect Themselves in Different Ways

Living things are often in danger of being eaten. Many animals eat plants. **Predators** hunt and eat other animals. Living things protect themselves from danger in lots of different ways. These differences show variation.

In times of danger, sea cucumbers can shoot out their sticky guts! They grow the guts back later.

**Camouflage** helps this flatfish hide from predators. The color of the fish matches the color of the sand.

Coral snakes can kill bigger animals with one bite. The bright stripes on a coral snake warn predators away.

Boxwood leaves smell bad to deer. Deer won't eat them.

**Mimicry** protects milk snakes from predators. These harmless snakes look a lot like dangerous coral snakes, so predators stay away from the milk snakes, too.

It's easy to observe how living things are different. Still, living things are the same in important ways, too. Noticing how living things are similar can be harder—but it is very useful . . .

# How Living Things Are Similar

Different species are **related** to one another. Scientists who study a species often ask: Which other species is this species closely related to? To learn how closely related different species are, scientists use many kinds of **evidence**. They get some of their evidence by observing the characteristics of species. When two species share a characteristic, it may be evidence that they are related.

Still, different species may share some characteristics without being closely related. For example, a red bird and a red flower share the characteristic of being red. This does not mean that they are closely related! Scientists collect lots of evidence in order to understand which species are closely related.

wolf

fox

wild dog

There is a lot of evidence that these animals are all closely related to one another. Can you observe some characteristics they share?

Look at the two living things below. Which one is more closely related to the red bird above?

Scientists **identify** groups of related living things. One group is plants. Almost all plants can make their own food using sunlight. Another group is animals. Animals eat plants, other animals, or both plants and animals. These very big groups are made up of smaller groups. One smaller group is all the animals that have four **limbs**, such as legs.

# Some Living Things Have Four Limbs

Animals with four limbs are all related. Even though a lion might not look like a frog, these animals share the characteristic of having four limbs. A human and a rat are also similar in this way. The fact that all these animals have four limbs is one piece of evidence scientists use to understand that they are all related.

A bird's wing is a kind of limb, too. Birds have four limbs: two legs and two wings. The wing of a bird has the same basic bones as your arm. These limb bones are evidence that birds belong in the same group as lions, frogs, rats, humans, and other animals with four limbs.

Herons have four limbs: two wings and two legs. They use their long legs to wade in the water, where they hunt for fish.

Sloths have limbs with long claws that make it easy for them to hang upside down.

Frogs have four limbs. This one is using its strong back legs to swim.

Geckos have four limbs with toes that stick to walls and trees.

This group is smaller than the group that includes all animals, but it is still a big group. The group of animals with four limbs contains many smaller groups . . .

15

# Some Living Things Have Feathers

A smaller, more closely related group is made up of birds. All birds have feathers to cover them. This shared characteristic is one piece of evidence showing that all birds are related.

Birds do not all look alike, though. Each species of bird has feathers that make it look different from all other birds. Colorful feathers help some birds attract mates. Dull brown and green feathers help other birds hide in trees.

16

Frogmouth birds have feathers that are almost like whiskers. These feathers help the birds feel for tiny animals to eat.

Owls have wing feathers that let them fly silently. They can swoop down and catch mice by surprise.

Ducks have feathers that keep them warm and dry in cold water.

Animals aren't the only living things that can be grouped together . . .

17

# Some Living Things Have Flowers

Scientists identify related groups of plants, too. The plants in each group share important characteristics. For example, some plants have flowers, and others do not. All the plants that have flowers are related.

Flowers can be very different. They come in all different sizes, shapes, colors, and smells. Even with all this variation, flowering plants are grouped together. The shared characteristic of having flowers is evidence that these plants are all related.

The bright yellow color of this black-eyed Susan helps bees find it.

18

Flowers have dusty **pollen** inside, and they often have sweet **nectar**, too. Small animals visit flowers to drink the nectar, and the animals spread pollen from flower to flower. Spreading pollen is called **pollination**. Pollination helps the plants make fruit filled with seeds that can grow into new plants. Many flowers have bright colors, strong smells, or shapes that attract animals such as bees, birds, flies, and even bats.

This bee is pollinating a flower. You can see the dusty yellow pollen on the bee's body.

Bats pollinate this kind of cactus. The cactus flowers have a sweet smell that helps bats find them at night.

The largest flowers in the world smell like rotting meat. This smell attracts flies that pollinate the flowers.

There are lots of different groups of plants and animals. Still, all living things on Earth are similar in at least one way . . .

# All Living Things Are Related

There is one way that *all* living things are the same—they are all made of tiny living parts called **cells**. Some living things have only one cell. These living things are almost always very small. Other living things are made of a great many tiny cells.

Whether they are big or small, simple or **complex**, plant or animal, all living things are made of tiny cells. Cells are the basic building blocks of living things. The shared characteristic of being made of cells is evidence that all living things are related to one another.

This is a buttercup plant shown under a microscope. You can see the tiny cells that make up the plant.

20

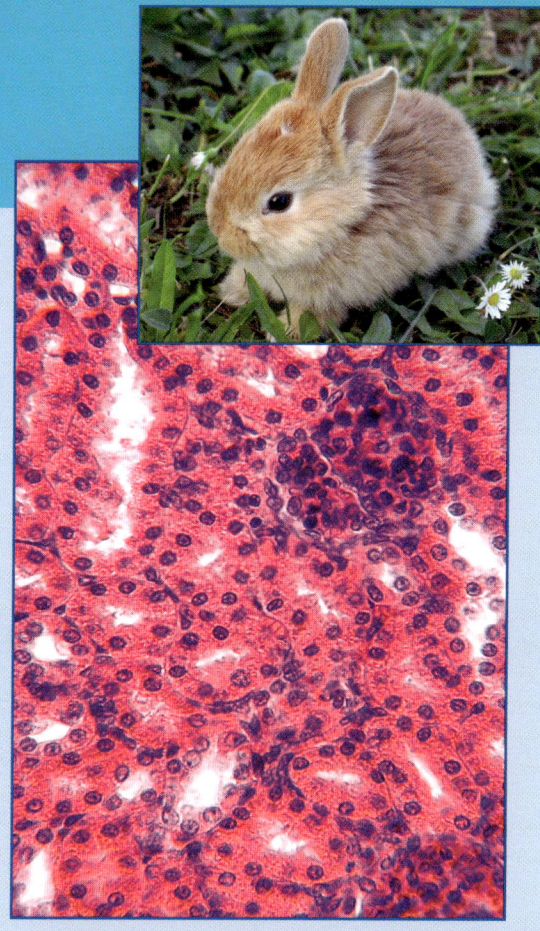
Rabbits are also made up of tiny cells. These are cells from a rabbit shown under a microscope.

You are made up of cells, too. These are human fat cells shown under a microscope.

This whole living thing is one tiny cell! It is much too small to see without a microscope.

21

The living things on these pages are examples of the wonderful variation we can see on Earth. How are these pairs of living things different? What characteristics do they share?

fern

cactus

tree frog

warthog

ant

mantis

buttercup

blue whale

23

# Glossary

**camouflage:** color or markings that help a living thing hide by blending in with the things around it

**cells:** tiny parts that make up living things

**characteristic:** something you can observe about a thing, such as how it looks or what it does

**compare:** to notice how two or more things are alike or different

**complex:** complicated, having many parts

**evidence:** clues that help explain something or answer a question

**identify:** to figure out what something is or the group it belongs to

**limb:** a body part, such as an arm or a leg, used for holding things or moving

**microscope:** a tool that makes very small things look bigger

**mimicry:** when a living thing is protected by looking or acting like another living thing

**nectar:** sweet liquid made in flowers

**observe:** to use any of the five senses to gather information about something

**pollen:** a dusty or sticky powder that is found in flowers and helps make seeds

**pollination:** the spreading of pollen from flower to flower

**predator:** an animal that hunts and eats other animals

**related:** coming from the same family or group of living things

**species:** a group of living things that are more closely related to one another than to any other living things

**variation:** differences in living things